T0325129

The Whys of a Scientific Life

Global Science Education

Professor Ali Eftekhari
Series Editor

Learning about the scientific education systems in the global context is of utmost importance now for two reasons. Firstly, the academic community is now international. It is no longer limited to top universities, as the mobility of staff and students is very common even in remote places. Secondly, education systems need to continually evolve in order to cope with the market demand. Contrary to the past when the pioneering countries were the most innovative ones, now emerging economies are more eager to push the boundaries of innovative education. Here, an overall picture of the whole field is provided. Moreover, the entire collection is indeed an encyclopedia of science education, and can be used as a resource for global education.

Series List:

The Whys of a Scientific Life
John R. Helliwell

The Whys of a Scientific Life

John R. Helliwell

CRC Press
Taylor & Francis Group
Boca Raton London New York

CRC Press is an imprint of the
Taylor & Francis Group, an **informa** business

CRC Press
Taylor & Francis Group
6000 Broken Sound Parkway NW, Suite 300
Boca Raton, FL 33487-2742

International Standard Book Number-13: 978-1-138-38979-3 (Hardback)

Visit the Taylor & Francis Web site at
http://www.taylorandfrancis.com

and the CRC Press Web site at
http://www.crcpress.com

It is important that we reflect upon our craft, since our understanding of science will inform public policy towards it—"science policy" as it is called.

John Polanyi, winner of the Nobel Prize in Chemistry

Contents

Series Preface

CONTRARY TO THE COMMON perception, the concept of education is not straightforward; both its purpose and its methods are subject to controversies. Plato was among the first who attempted to articulate the foundation of education by putting an emphasis on disciplines, which can directly help us to understand the universe.

Science education should be revisited in the changing climate and emerging needs of today. Both the concept and methods of education have been considerably evolved in the digital era. In the highly competitive market of higher education, higher-education institutions (HEIs) are too conservative to implement innovative changes. This series attempts to provide practical perspectives on various aspects of science education. In this book, John Helliwell takes us on a journey to the fundamental roots of science. With a long and fruitful experience in academia, he tells us where we are, why we are here, and how to survive.

The world is now more connected, and we can learn from a number of different systems, although direct interactions might be challenging due to social and cultural barriers. This series aims to provide a medium for a broad range of audience to close this gap. I might add that the authors of this series do not necessarily share my opinions or concerns, of course.

The inception of this book series harks back to a friendly conversation with Hilary LaFoe, senior acquisitions editor at CRC Press. I should give at least half of the credit to her because the initial idea was hers. I would like to thank her because this project could not have materialised without her sincere commitment. I also encourage readers to give us feedback in order to help us take a small step towards furthering the concept of education at all levels according to the changing climate of globalisation.

Dr. Ali Eftekhari
Global Science Education

Preface

A S A READER, TO open a new book for which one finds the title interesting stimulates an exciting anticipation. To start writing a new book is also exciting, but it is also a challenge, especially with this one. I think that it is a good time for me to share my thoughts on this topic based on my 40 years of experience in scientific research. It is an intimate thing to share thoughts with you, the reader, about why I have undertaken the various themes of the research that I have done and continue to do to this day. I also wish to highlight examples from the history of science spanning the centuries. The public understanding of science and its discoveries needs clarity from scientists, especially what we are about and why. This is my contribution. I also hope that schoolchildren interested in science will find my book interesting, as will their parents, who naturally are following the interests of their children's education with great care.

In my previous book, *Skills for a Scientific Life*, I focussed on how to approach the many diverse tasks that make up a life in science. The skills involved in how to do things can be shaped and improved. The challenge is very much with scientists to focus on their skills. That book contained a substantial chapter on the subject of ethics in science and of the scientist, and it provided guidance on how to proceed ethically.

This new book, *The Whys of a Scientific Life*, concentrates much more on the environment of the scientist and the choices that must be made in response to the myriad of opportunities to apply one's scientific skills, both in general life and in the research laboratory. At its core, scientists ask questions based on curiosity. So they can apportion precious time out of each working day to personal research that addresses questions that they see as important. One's environment offers a variety of opportunities. One needs to secure research funding in order to have a well-equipped research laboratory, with its trained staff, equipment, and consumables. For specific research and discovery projects at any one time, there will be research assistants funded by research agencies. There also is the potential of engaging in collaborations with other researchers to tackle scientific challenges of a larger scope.

Furthermore, such a research environment allows for the education of upcoming scientists (that is, students). At the University of Manchester these last 30 years, as a professor of structural chemistry, often as a joint appointment scientist with the nearby Daresbury Laboratory, with the European Synchrotron Research Facility and the Institut Laue Langevin in Grenoble, and most recently with the Diamond Light Source at the Rutherford Appleton Laboratory, I have felt very fortunate to have had the chance to work in such a fertile research environment. I hope that my case studies from this background will be of educational interest, and in some way assist scientists to make good choices, and also allow the public (including schoolchildren) to understand better the "whys" of a scientific life.

In summary, this relatively short book dissects the various reasons why we undertake science and discovery. I provide case studies from my career, which has given me a wide range of insights into the whys of science and discovery. I describe science and discovery in simple terms, mainly free of jargon, based on my outreach activities to the public and to schoolchildren. And I end with a chapter on the joys of a scientific life—what better place to conclude than that!

In one sentence: This book describes why we undertake science and discovery, in various forms, and explores the possible limits to discovery to be faced jointly by scientists and society in general.

Now, read on...

Acknowledgements

I COMMEND PROFESSOR ALI EFTEKHARI as editor, for launching this new series on Global Science Education. I record here my pleasure at supporting the initiative with this book, which Hilary LaFoe, my CRC Press Taylor and Francis publisher contact, invited me to contribute. The instructions to authors are clear, which I acknowledge, and state that the series can cover the following topics:

Titles published in the CRC Press Focus program could include:

- A short overview of an emerging area or "hot topic."

- A detailed case-study.

- Reworking of research for a policy audience.

- Analytical or theoretical innovations.

- A timely response to current affairs or policy debates.

- Information and analysis for professionals and practitioners within a certain field.

- Edited collection with a variety of viewpoints.

My book is relevant to several of these aspects. I share here the answers of Professor Eftekhari to two of the questions for the assessment of my proposal that he had to complete, which I also appreciate and acknowledge:

Is there a need for this book? If so, what problems will it help readers to solve?

Definitely yes. We surely need more books with this outlook, as the essence of science has been somehow obscured by the rush for technological development.

What do you consider to be the primary and secondary markets for this proposal (e.g. undergraduate, graduate, academia, industry.)?

The market should be quite broad for this book. Not only discipline-wise, but also I believe the audience will be at different levels of education.

I thank all my students, staff, and collaborators throughout my scientific research career for their contributions to the research themes I describe in this book, which I have chosen as illustrations of why we did the things that we did. I am very grateful to the rich and diverse environments that have shaped my research and my career development. These include the University of Manchester (1989–) and the Daresbury Laboratory (1976–2008), to both of which I am especially grateful. I also thank the large number of agencies that have funded my research proposals. I have been involved in the International Union of Crystallography (IUCr) in numerous capacities, which has greatly diversified my experience, including representing IUCr at the International Council for Scientific and Technical Information and at the International Council for Science Committee on Data. These organisations have brought me into direct contact with scientists from a diverse range of science subjects and which I greatly appreciate.

I am grateful to various colleagues who have commented on my book manuscript, Dr Alice Brink of the School of Chemistry at the University of the Free State, Bloemfontein, South Africa; Emeritus Professor Carl Schwalbe, of the Department of Pharmacy, Aston University, United Kingdom; and Katrine Bazeley, member of the Royal College of Veterinary Surgeons.

About the Author

John R. Helliwell is Emeritus Professor of Chemistry at the University of Manchester. He was awarded a DSc degree in physics from the University of York in 1996 and a DPhil in molecular biophysics from the University of Oxford in 1978. He is a Fellow of the Institute of Physics, the Royal Society of Chemistry, the Royal Society of Biology, and the American Crystallographic Association. In 1997, he was made an Honorary Member of the National Institute of Chemistry, Slovenia. He was elected a corresponding member of the Royal Academy of Sciences and Arts of Barcelona, Spain, in 2015. He was made an Honorary Member of the British Biophysical Society in 2017. The same year, he became a Faculty 1000 Member, charged with highlighting significant science publications. He was a Lonsdale Lecturer of the British Crystallographic Association in 2011, the Patterson Prize Awardee of the American Crystallographic Association in 2014, and the Max Perutz Prize Awardee of the European Crystallographic Association in 2015.

CHAPTER 1

Introduction to "Because"

A NY QUESTION "WHY" IS basically answered by giving the following reasons. I follow this simple rule for most of my chapters (that is, where there are clear reasons).

A scientific research publication can be classified according to its roles in science and discovery. In this way F1000Prime (see https://f1000.com/prime/my/about/evaluating) allows its faculty to evaluate such roles. These are, and I quote from the F1000Prime website, as of 2018:

Assign (publication) classifications

- Confirmation: validates previously published data or hypotheses
- Controversial: challenges established dogma
- General Interest
- Good for Teaching: key article in field and/or well written
- Interesting Hypothesis: presents new model
- New Finding: presents original data

- Novel Drug Target: suggests new targets for drug discovery

- Refutation: disproves previously published data or hypotheses

- Technical Advance: introduces a new practical/theoretical technique, or novel use of an existing technique

- Negative/Null results: particularly for clinical trial–related articles.

If one considers examples of well-known, major discoveries, we can assign them to be one or more of the following types:

- The discovery of the double helical structure of DNA in 1953 by Watson and Crick, based on DNA fibre X-ray diffraction data of Rosalind Franklin, and a mathematical formalism for helical fibre diffraction, gave a direct idea of the molecular basis of heredity. Later, it gave the genetic basis of some medical pathologies, introduced genetic fingerprinting for forensic science, and most recently genetic editing for curing some diseases. The double helix discovery itself was a curiosity driven piece of science. It was preceded by research at the University of Leeds by Florence Bell with Bill Astbury in 1939 and Maurice Wilkins at Kings College. Figure 1.1 is a picture showing a portion of the DNA double helix based on a crystal structure analysis I undertook with my coworkers in 1996.

- The discovery by Albert Einstein of the equation $E = mc^2$ was again curiosity driven. It led to the use of nuclear fission as a peaceful source of energy on the one hand, and was employed to create the ultimate weapon of our age: the atomic bomb on the other.

- The discovery of the theory of evolution by Charles Darwin resulted from the collection of natural examples of species during his ocean voyage to South America. During his exploration, did Darwin predict that he would go on to hypothesize the theory of evolution? I don't think so.

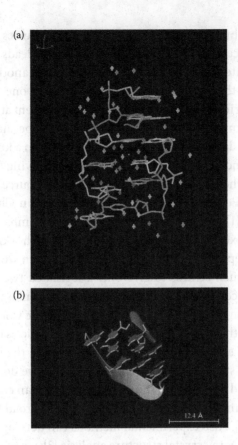

FIGURE 1.1 (a) The double helical structure of DNA as illustrated by a short portion 'oligonucleotide' with the double helix axis vertical. One vertical nucleotide strand is visible to the left and its paired one vertical to the right. The + symbols depict bound water molecules. Based on the crystal structure held at the Protein Data Bank entry 242D based on the research of M.R. Peterson, S.J. Harrop, S.M. McSweeney, G.A. Leonard, A.W. Thompson, W.N. Hunter and J.R. Helliwell "MAD phasing strategies explored with a brominated oligonucleotide crystal at 1.65Å resolution" (1996) *J. Synchrotron Rad.* 3, 24–34. This figure was prepared in Coot (Emsley, P., Lohkamp, B., Scott, W. G. & Cowtan, K. (2010). *Acta Cryst.* D66, 486–501.) (b) This is a different representation shown to highlight the nucleotide bases pairing; the double helix axis runs from top left to bottom right. This figure was prepared in CCP4MG (McNicholas, S., Potterton, E., Wilson, K. S. & Noble, M. E. M. (2011) *Acta Cryst.* D67, 386–394.)

In these three examples from the last two centuries, we can see that the process of science is ongoing; one thing leads to another. They illustrate how one development builds upon another, and the reason why the next step is done changes from one category to another. Curiosity is, in my view, most often present at the start of new opportunities. We don't invent a lightbulb because we need to see in the dark; rather, we understand that from a knowledge of electricity and an experiment that leads to a glowing (that is hot) wire, it has the potential to be developed into a source of light.

As John Polanyi, winner of the Nobel Prize in Chemistry in 1986, put it [1]: "[In my country, we] have, for example, numerous 'Centres of Excellence' because we recognize that the skill on which discovery depends is possessed by a few. But then we proceed in evaluating such centres to give only a legislated twenty percent weight to 'excellence.' A preposterous eighty percent is reserved for considerations having to do with 'socio-economic worth'."

So we see that there are strong opinions and policies that fashion the science research environment, and therefore the reason why we as scientists do, or must do, the things that we do, both to be ourselves and to obtain funding so our research can continue. It is also important to remember that it can be a long road to a science discovery [2]. Also experimental methods can develop over many decades e.g. X-ray crystal structure analysis [3].

REFERENCES

1. J. Polanyi. (1986). *On Being a Scientist: A Personal View.* Accessed August 9, 2018, from https://www.nobelprize.org/nobel_prizes/chemistry/laureates/1986/polanyi-article.html.
2. Kirsten T. Hall. (2014)."The man in the Monkeynut Coat: William Astbury and the Forgotten Road to the Double-Helix." Published by Oxford University Press.
3. William Lawrence Bragg. (1975). "The Development of X-ray Analysis." Published by Dover Press, Mineola, New York.

I

Fundamental Science

CHAPTER 2

Because We Ask a Question

ARGUABLY, THE MOST FAMOUS example of a significant question is Albert Einstein asking, "What if the speed of light is finite?" We don't know why he asked that question, but the mathematical physics that flowed from it via the equations he developed led to new understanding of physical phenomena, such as the increase of mass as a particle is accelerated to near that speed (see Eq. 2.1).

$$m = m_0/(1 - v^2/c^2)^{1/2} \qquad (2.1)$$

This is Einstein's relativistic mass equation, where m is the relativistic mass at the particle's velocity v, m_0 is the particle's rest mass, and c is the speed of light, which is a constant.

For the research fields that I have chosen to work in over the years of my career, a major influence on me happened in my final year as an undergraduate physicist, when I took an option course on biophysics. A recommended reference appeared in *New Scientist* on the 17th June 1971 [1], written by Dr Max Perutz (1914–2002), entitled "Haemoglobin: the molecular lung," which

I read with awe and amazement. He asked how oxygen gets from the lung to the muscle for the muscle to do its work. To address this question, Perutz grew his own crystals of haemoglobin and applied the methods of the science of crystallography to determine its molecular structure. A key feature, he found, was that the iron atom in the haem moved into the haem plane upon oxygen binding to it on one side.

The science of crystallography gradually developed to allow for making an error estimate on the movement distance of the iron atom, so as to be really sure that it was a significant distance. However, Max Perutz saw bigger changes in the haemoglobin protein structure arising from the first oxygen binding, making it easier for three more oxygens to bind. The structural processes involved were structurally *cooperative*. This is an example of curiosity-driven science.

Max Perutz was employed at the Medical Research Council's Laboratory of Molecular Biology in Cambridge, initially formed as an offshoot of the University of Cambridge's Physics department, the Cavendish Laboratory. The Charter of the Medical Research Council [2] defines its missions as follows:

> The objects for which the Council is established and incorporated are:
>
> a. To promote and support, by any means, high-quality basic, strategic and applied research and related post-graduate training in the biomedical and other sciences, with the aim of maintaining and improving human health;
>
> b. To advance knowledge and technology (including the promotion and exploitation of research outcomes), and provide trained researchers, which meet the needs of users and beneficiaries (including the providers of health care, and the biotechnology, food, health-care,

medical instrumentation, pharmaceutical and other biomedical-related industries), thereby contributing to the maintenance and improvement of human health, the economic competitiveness of our United Kingdom, and the quality of life;

c. In relation to the activities as engaged in by the Council under (a) and (b) above and in such manner as the Council may see fit:

i. To generate public awareness;

ii. To communicate research outcomes;

iii. To encourage public engagement and dialogue;

iv. To disseminate knowledge; and

v. To provide advice.

Through Max Perutz's basic science discoveries on the structure and function of haemoglobin, we can see immediately the hope of the Medical Research Council that such basic science work would ultimately be "contributing to the maintenance and improvement of human health." Indeed, the haemoglobin X-ray crystal structure allowed a variety of insights into the molecular basis of blood pathologies because amino acid mutations that were known to be associated with different pathological conditions could now be mapped directly onto the surface of the haemoglobin's three-dimensional (3D) structure. Overall, one can see in this example the happy concurrence of the basic science question being posed by a scientist and the view of his employer, the Medical Research Council. Max Perutz's early funding, however, was not from the Medical Research Council [3]. This was a grant from the Rockefeller Foundation until 1945 and he was then awarded an Imperial Chemical Industries Research Fellowship.

REFERENCES

1. Max Perutz (1971). "Haemoglobin: the molecular lung" New Scientist 17th June 1971, 676–678.
2. Charter of the Medical Research Council. Accessed August 9, 2018, from https://www.mrc.ac.uk/documents/pdf/mrc-charter/.
3. MRC Laboratory of Molecular Biology. http://www2.mrc-lmb.cam.ac.uk/about-lmb/archive-and-alumni/alumni/max-perutz-1914-2002/

CHAPTER 3

Because We Make a Hypothesis

SCIENCE CAN MOVE FORWARD when a scientist makes a hypothesis. This impinges on the philosophy of science and on the reproducibility or irreproducibility of science. An aspect of science that is close to asking a question is making a hypothesis. A great thinker on the way science works and moves forward was Karl Popper. He described the importance of a hypothesis as follows: If one wishes to prove the hypothesis that all swans are white, finding one white swan after another is insufficient. Rather, it is obviously important to refute the hypothesis that all swans are white by finding a black swan. So science can move forward not by being reproducible, but by being *irreproducible*. Today, the notion of irreproducibility of science is seen as a major concern, and indeed it is especially important if research is done poorly or fraudulently. The antidote to poor research or fraudulent research is to insist that publications be accompanied by the scientists' data, as well as clear metadata (that is, the descriptions and numerical core information describing the measured data). As the Karl Popper example shows, the data of observation of a black swan is of a very simple type, just involving counting, but nevertheless it shows how irreproducibility that is firmly established is an important way for science to move forward.

In physics, the Michelson–Morley experiment sought to prove the hypothesis that there was an ether through which light moved (that is, light as a wave had to move through something, not through a vacuum). Thus, they argued, the speed of light would vary according to whether the Earth (the laboratory light source) moved into the ether or away from it. Michelson and Morley's findings, published in 1887, showed that the speed of light was constant, regardless of the direction of travel [1]. This result was very important. A leading physicist of the age, Lord Kelvin, said that it was one of two aspects [2] that impeded the understanding by physics of the universe, which he portrayed as "the beauty and clearness of the dynamical theory, which asserts heat and light to be modes of motion, is at present obscured by two clouds. I. The first came into existence with the undulatory theory of light, and was dealt with by Fresnel and Dr Thomas Young; it involved the question, How could the earth move through an elastic solid, such as essentially is the luminiferous ether? II. The second is the Maxwell-Boltzmann doctrine regarding the partition of energy." This speech, together with other comments attributed to Lord Kelvin (such as by the physicist Albert Michelson in an 1894 speech), indicate that he strongly believed that the main role of physics at that time was simply to measure known quantities to a great degree of precision, out to many decimal places of accuracy.

REFERENCES

1. A. A. Michelson and E. W. Morley. (1887). "On the relative motion of the earth and the luminiferous ether." *American Journal of Science, 3rd Series*, Volume XXXIV, 333–345. [Available via https://www.thoughtco.com/the-michelson-morley-experiment-2699379.]
2. Right. Hon. Lord Kelvin G.C.V.O. D.C.L. LL.D. F.R.S. M.R.I. (1901) I. Nineteenth century cloud sover the dynamical theory of heat and light." The London, Edinburgh, and Dublin Philosophical Magazine and Journal of Science." 2:7, 1–40, doi: 10.1080/14786440109462664.

CHAPTER 4

Because We Wish to Make a Collection

THERE ARE SITUATIONS WHERE we deem systematically gathering a collection as a worthwhile scientific pursuit. The sequencing of every gene of the human genome, where one gene codes for and is expressed as one protein, is one such project. Both public funding agencies and commercial enterprises agreed (and indeed, even competed with each other) to pursue this goal. The public funding agencies were motivated because their leaders thought that achieving this would take the understanding of health and disease to a whole new level, and with it the chances for devising new cures.

By contrast, the thinking of commercial enterprise was that important genes could be identified and patented. This became a controversy that was eventually resolved in the courts; it proved incorrect that patents on genes could be taken out and commercialised. To many of us scientists, this was the obvious conclusion. In fact, it should have been the starting premise that it should be impossible to patent any part of our human genetic inheritance. However, it fell to the courts to make this ruling.

The collection of gene sequences in the human genome is an example of scientific method where there was no real hypothesis, other than that humans are bound to learn something from the human genome research that we perform; this included advancing basic science (namely, human biochemistry), and the findings were applied to the development of new medicines. A very famous historical example of the power of a systematic collection was that of Charles Darwin, whose observations of the geographical distribution of wildlife and his collection of fossils led to his theory of biological evolution of species. He undertook his research during his five-year voyage on the *HMS Beagle* to South America (1831–1836). In his subsequent book *On the Origin of Species*, (1859) [1], he stated; "As many more individuals of each species are born than can possibly survive; and as, consequently, there is a frequently recurring struggle for existence, it follows that any being, if it vary however slightly in any manner profitable to itself, under the complex and sometimes varying conditions of life, will have a better chance of surviving, and thus be naturally selected. From the strong principle of inheritance, any selected variety will tend to propagate its new and modified form… (and) from so simple a beginning endless forms most beautiful and most wonderful have been, and are being, evolved." Darwin's theory of evolution was published, along with his independently arrived at theory, by Alfred Russell Wallace in *On the Law which has Regulated the Introduction of New Species* (1858) [2].

A recent proposal is to gather together a collection of all the deoxyribonucleic acid (DNA) gene sequences of each species [3]. It is known as the Earth BioGenome Project, and it is akin to a modern Noah's Ark. The central goal of the Earth BioGenome Project is to "understand the evolution and organization of life on our planet by sequencing and functionally annotating the genomes of 1.5 million known species of eukaryotes, a massive group that includes plants, animals, fungi and other organisms whose cells have a nucleus that houses their chromosomal DNA. To date, the genomes of less than 0.2 percent of eukaryotic species have been sequenced... The completed project is expected to require about one exabyte (one billion gigabytes) of digital storage capacity. [4]" The authors of the proposal [3] describe it as "a moonshot for biology that aims to sequence, catalog, and characterize the genomes of all of Earth's eukaryotic biodiversity over a period of 10 years. The outcomes of the Earth BioGenome Project will inform a broad range of major issues facing humanity, such as the impact of climate change on biodiversity, the conservation of endangered species and ecosystems, and the preservation and enhancement of ecosystem services." I would imagine that Darwin would have been a strong supporter of the proposal.

REFERENCES

1. C. Darwin. (1859). *On the Origin of Species*. London: John Murray.
2. C. Darwin, and A. R. Wallace. (1858). "On the tendency of species to form varieties; and on the perpetuation of varieties and species by natural means of selection". *J. Proc. Linn. Soc. Lond. Zool* 3(9): 46–50.
3. H. A. Lewin, G. E. Robertson, W. J. Kress, et al. (2018). "Earth BioGenome project: Sequencing life for the future of life". *PNAS* 115(17), 4325–4333. www.pnas.org/cgi/doi/10.1073/pnas.1720115115.
4. Phys.org. *Earth Biogenome Project Aims to Sequence DNA from All Complex Life on Earth*. Accessed August 9, 2018, from https://phys.org/news/2018-04-earth-biogenome-aims-sequence-dna.html.

CHAPTER 5

Because of "What Happens If?"

NOT AS STRONG AS making a hypothesis, we have situations where we wonder "what happens if" we try something different. It may simply be a good guess—what I call a hunch. In my own field of crystallography, where I determine crystal structures using X-rays and neutrons, we naturally have the need for crystals. The science of crystallisation is a research field in its own right, and it is a multiparameter challenge to determine the chemical conditions to grow the best crystal from a liquid or gel [1].

In the process of crystal growth, the first step is promoting nucleation (i.e., the coming-together of a small group of molecules). This is followed by a second step: facilitating crystal growth. The molecules being crystallised come from the solution, whose physical properties are thereby altered. One such property is the density of the liquid, and the different density values through the solution lead to movements of the fluid in reaction to the Earth's gravity. Another effect is that a crystal under its weight can sediment (i.e., fall to the bottom of the solution). Both effects can be eliminated if there is no gravity, of course. So, what happens if gravity is eliminated?

Of course, as I have explained it, there is a hypothesis behind the idea, and calculations can be performed to predict the fluid flow under various fluid geometries. One such geometry is best suited to exploit a zero gravity environment, liquid diffusion. In practice, on board the space shuttle (not infinitely far from the Earth, obviously), the gravity isn't quite zero—in fact, it is microgravity, one-millionth of the gravity on the surface of the Earth.

Predictions are one thing, but one simply has to try them to see what happens to the crystal growth process, and indeed, whether the perfection of the crystal is better. My contribution to this project came in the evaluating of crystal quality and perfection. Suffice to say, crystals could be grown whose perfection I tested by the geometric index of their mosaicity; this is a measure of how well every molecule is packed into a growing crystal, with better being a lowered mosaicity of a crystal. It took quite a while to prove the effect, as there were precise controls to be set up on Earth parallel to those on the shuttle. In addition, the best geometry of liquid-liquid diffusion has taken a while to establish. There were also temporary gravity disturbances of the astronauts taking their daily exercise, which created vibrations akin to gravity. Overall, accessing the environment of the shuttle is not very convenient, and so other methods were developed on Earth to mimic as closely as possible the shuttle's microgravity conditions. These included crystallisation in gels to prevent fluid flow or crystal sedimentation. Miniaturising the crystal growth sample chamber, even as small as nanolitre volumes, was another such mimic of microgravity, because the small volume also prevents fluid flow and sedimentation. All in all, the science of crystallisation was advanced as the community of crystal growers and crystallographers like myself undertook research within the new environment of microgravity, harnessing its advantages as well as possible.

REFERENCE

1. N. E. Chayen, J. R. Helliwell, and E. H. Snell. (2010). *"Macromolecular Crystallization and Crystal Perfection" (International Union of Crystallography Monographs on Crystallography)*. Oxford University Press, International Union of Crystallography Monographs on Crystallography.

CHAPTER 6

Because One Thing Leads to Another

I T IS A WELL-KNOWN saying that one scientific discovery leads to many another, or indeed many new questions. Not all those new questions are at the highest priority for investigation, but generally at least one is. Usually, the researchers who publish a study carry out follow-up research if time and funds permit. Naturally, one hopes that in any case, other researchers will build on what one does—a phenomenon referred to as *research impact*. An unexpectedly popular piece of research that my laboratory and my collaborators undertook was the study of the protein complex that caused the particular colouration of the European lobster (*Homarus gammarus*), which has a distinctive blue-black colour—at least when it is uncooked, because of course upon being cooked, the lobster turns red. Our contribution was to work out the three-dimensional (3D) structure— the atomic layout—of the crustacyanin protein (see Figure 6.1). This protein is a dimer holding two carotenoid molecules, known as *astaxanthins*. Carotenoids are a general class of organic molecules, and they are coloured pigments (generally red, orange, or yellow) that give biological materials such as carrots their colour.

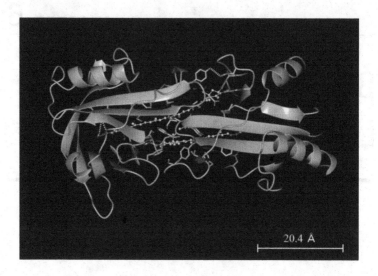

FIGURE 6.1 The X-ray crystal structure of β-crustacyanin [1]. This protein is a dimer holding two astaxanthin carotenoid molecules. This graphic shows a simplified model of the protein polypeptide chain in blue (the so-called ribbon diagram format), and in the centre, the two astaxanthin carotenoids depicted in green. This graphic is based on M. Cianci et al. (2002); it was prepared in CCP4MG, a molecular graphics computer program. (McNicholas, S. et al. *Acta Cryst.* 2011, D67, 386–394. [2])

The crystals were beautiful, having a distinct blue colour; they were grown by our collaborator, Prof Naomi Chayen at Imperial College in London, a world-ranked expert on protein crystallisation. The distinct way that the protein held the two astaxanthins revealed in our crystal structure suggested several ways in which the colour change of the free astaxanthin from orange-red at that dilution to blue occurred when held in the protein. Of course, we could readily imagine that, upon cooking, the bound astaxanthins are released from the protein's grip.

Our article [1] led to numerous other scientific studies, and it has been well cited (in fact 150 other publications have referred to it). The scientific fields of the publications that have cited our study are very diverse, ranging from biophysics, chemistry, marine biology, and food science. For ourselves, we wished to move on to study the full crustacyanin complex because so far, we had studied a one-eighth fraction of a super-big complex. Crystals could be grown, but unfortunately, they were insufficiently ordered to allow us to determine the atomic layout. This we had hoped would allow us to understand why the blue β-crustacyanin crystal was different from the distinctly darker blue α-crustacyanin crystal. These colours were quantified in their wavelengths by an instrument called a spectrometer. The colour shift in the β-crustacyanin crystal from orange-red to blue was 100 nm, and the further spectral shift to the darker blue α-crustacyanin crystal was another 50 nm. "How did this occur?" was the next scientific question that we wished to answer. We turned to the method of electron microscopy and determined a layout of the blue α-crustacyanin protein, but at insufficient resolution to see individual atoms as we had done with X-ray crystallography.

Would that be the end of this research theme for me? Actually, no, because the Australian lobster (*Panulirus Cygnus*) and the South African lobster (*Homarinus capensis*) are both red when they are alive, and so one asks why. Also, there is a purple starfish (*Linckia laevigata*) [3] that has a protein complex known as *linckiacyanin*, which we imagine will be similar to the α-crustacyanin, but different because of its different colour. The wider aspects of this research involve questions like "Why has nature evolved these different colours for these marine organisms?" "What predator prey relationships might exist to lead to the current situations in different parts of the world for these marine organisms?" "Is climate involved?" "Is their preferred depth where they live in the oceans important?"

REFERENCES

1. M. Cianci, P. J. Rizkallah, A. Olczak, et al. (2002). "The molecular basis of the coloration mechanism in lobster shell: β-crustacyanin at 3.2 Å resolution." *PNAS USA* 99, 9795–9800.

2. S. McNicholas, E. Potterton, K. S. Wilson, M. E. M. Noble. (2011). "Presenting your structures: the CCP4mg molecular-graphics software." *Acta Cryst*, D67, 386–394.

3. S. T. Williams. (2000). "Species boundaries in the starfish genus Linckia." *Marine Biology*, 136(1), 137–148.

Because We Get Criticized

T HE PUBLICATION OF SCIENCE research is likely to be a start of future research, both by one's own laboratory and others. If the research is underpinned by new data, which will be the case if it is an experimental study rather than a theoretical or modelling one, then the data will be or should be *FAIR*, a simple label used by the data science community to describe all that is important about scientific data—they should be *findable*, *accessible*, *interoperable*, and *reusable*. Currently, this is an objective in research data management.

The advantage of one's research data being FAIR is that others can make their own calculations with it. This, of course, means that the science research that one has done is reproducible directly by the reader of a science publication. There is currently a major concern about the irreproducibility of some research studies. This can mean that those studies get criticized, and rightly so if the research was done incorrectly or the data set size (the sample size, as statisticians call it) was too small to prove what the article claimed. Such cases should never have been published if the journal had evaluated the original article submission properly.

What exactly does proper evaluation mean? In my experience, many journals only provide the article (i.e. the words describing a study) to its referees, chosen by an editor. There are exceptions to this, though. One such exemplary journal, where underpinning data and an automatically generated, standardised data-checking report, are made available to referees along with each article, is *Acta Crystallographica Section C: Structural Chemistry*. In such a case, readers and users of the underpinning data can be certain that the very best effort has been made to ensure the correctness of the results that are published by the journal. An important exception, however, was discovered when two research laboratories fraudulently edited their crystal structures by putting in a different metal than the one in an earlier, properly done, study. This came to light when the readership raised objections that the chemistry reported in those publications did not make sense—indeed was not even possible. Those fraudulent papers were openly criticised.

Because of that, the standardised data-checking program (i.e., software) was rewritten so that such crimes against science (and they were crimes) would be more easily detected. In all, around 100 fraudulent crystal structures were detected in articles published in a number of journals. As a result, all those articles were retracted from the journals concerned, and those researchers' employment with their universities in China was terminated. This story is an extreme, highly unusual example of when research is criticized. Of course, the data-checking software used by all those journals was improved, which was a good thing.

A different kind of criticism is where genuine mistakes occur. As I have mentioned, these should be detected by the journal editor and referees. A third area is where software is still in a developing state and new researchers (i.e. after publication) are applying the new software to existing scientific data, provided that it is FAIR. One example of this is when weak signals in the raw experimental data were not detected by the original software.

Thus, new software using the archived original data can extend a previous study's results with those data. That leads to a new article, which, inevitably has to criticise the old findings. It is possible that the weak signal in the data leads to a new result, correction of a previous result, or both. That would be a case of irreproducibility of research, which is a good thing—that is, science can advance by irreproducibility as well as reproducibility. A further aspect of such a case is when the original researchers look at what the new researchers have done and make yet further improvements and changes, again using the original data. This is science at its best. I described such examples in a keynote lecture that I delivered in August 2017, at the World Congress of Crystallography in Hyderabad, India [1].

Another point to make here is that as scientists, we need to remember that humility is a virtue. As Charles Darwin (1809–1882) put it in his autobiography [2]:

> I have steadily endeavoured to keep my mind free, so as to give up any hypothesis, however much beloved (and I cannot resist forming one on every subject), as soon as facts are shown to be opposed to it. Indeed I have had no choice but to act in this manner, for with the exception of the Coral Reefs, I cannot remember a single first-formed hypothesis which had not after a time to be given up or greatly modified.

I came across that quotation of Darwin in an essay by my longstanding collaborator, Professor Durward Cruickshank FRS (1924–2007) [3], where he added to Darwin's reflections as follows:

> The obligation of honesty is laid on the scientist. The actual standard of conduct in this is high for, in reading a scientific paper, one's last thought would be that there had been any deliberate deceit in the results presented.

REFERENCES

1. J. R. Helliwell. (2017). Keynote lecture at IUCr World Congress of Crystallography held in Hyderabad, India August 2017. Accessed August 8, 2018, from https://www.youtube.com/watch?v=lKx gvwNQNOk&list=PLbeFzl2kDyj9KKkQGAmaCwIvIOCXdG KxU&index=16.
2. C. G. Darwin. (1993). *Autobiography*. New York: W. W. Norton & Company. First published in 1887.
3. D. W. J. Cruickshank. (1965). "The universities and science." In the inset to the *University of Natal Gazette* XII(1), 1–8. Record of the address given at the graduation ceremony of the University of Natal in the City Hall, Durban, March 27, 1965.

CHAPTER 8

Because We Referee Other Scientists

W HY, AND WHEN, DO we criticise other scientists' work? Basically, we do this because very frequently, we get asked to referee other scientists' research, in the same way that we ourselves receive referees' reports to our own articles and to which we must respond to the handling editor. These requests come from numerous journals. One also receives a wide range of refereeing requests from science funding agencies (both national and international). This whole process is called *peer review,* which is an incredibly valuable process.

Peer review allows journal editors to be as certain as possible that a study is correctly done and the conclusions reached by the authors have a solid foundation before they publish an article. This assumes, of course, that the peer review has been done well. There are no surveys available to my knowledge as to whether any journals have done a proper evaluation of peer review of submitted articles and the data on which they rest. I know in my own field of crystallography that the only journal that reviews articles thoroughly is *Acta Crystallographica Section C,* published by the International Union of Crystallography, I mentioned this in Chapter 7 of this book. Publishers in general seem to me to be not worried enough about performing competent peer review of articles with data. For some years now, I have insisted on having the underpinning data, a standardised validation report from the relevant database, and the article. If I don't have these items, I do not provide my referee's report to the publisher, and I simply advise straight rejection of the submission.

Are funding agencies worried about journal peer reviews not being done properly for articles with data? It seems not. In fact, there is a trend to set up preprint archives on the model of arXiv, the physicists' repository for articles on the biosciences and chemistry. The agencies say that they like this approach because scientific results are shared earlier than by the conventional publishing route, which can delay results appearing for up to a year. The counterargument is that preprints are not peer reviewed and may be contain serious errors, which will mislead other scientists and waste scarce resources. Also, preprints do not always have the underpinning data available. There are good points on both sides of the argument, but for now, science journals are following the preprint approach, as physicists and physics journals have for many years with arXiv. A preprint can be criticised in forums such as social media, which can lead the authors to make improvements to their material before submitting it to a journal. Postpublication peer review, on the other hand, is rather haphazard. The challenge to preprint procedures is to achieve as good a peer-review standard as with journals (acknowledging that journals are not always perfect).

Peer review of submitted research proposals generally involves assessing the coherence of a research plan and the good sense of the chosen objectives. The planning aspect described in proposals has gradually and continually grown. Indeed, the proposers now must provide a timetable of how it will all unfold as the proposed research progresses. A proposal quite possibly will include a partial feasibility study in order to document that the work has good chances of success. The desire to have certainty that a research proposal will succeed risks that especially innovative research proposals will not secure funding.

My personal solution to this—namely, ensuring that my own "especially innovative" work is done—is to ensure that my own laboratory skills are up to the mark so that I can advance such work myself, ideally to the point of initial publication. I only realised that this was the best approach as I got older. Of course, it is said that one's most innovative research work is done whilst young. So, one way or another, good luck and chance contribute to the success of one's career as it progresses. The more systematic mentoring that is provided in universities to young scientists these days is a very useful development, as it offers early career scientists the opportunity to have conversations on such topics with a senior mentor, a role that I also undertook for several years.

CHAPTER 9

Because Something Unexpected Happens

A VERY WELL KNOWN EXAMPLE of this is how Alexander Fleming discovered penicillin. Apparently, when he returned from holiday in September 1928 to his laboratory in St Mary's Hospital, London, he was looking at petri dishes of the bacterium *Staphylococcus,* which causes boils, sore throats, and abscesses, and he saw one dish where a mould had grown. Around the boundary of the mould, there was a clear area. It was as if the mould had secreted something that had killed the bacteria there. The isolation of the active ingredient was difficult, and it was subsequently undertaken by Howard Florey and Ernst Chain at the University of Oxford. The pure compound was penicillin. The large-scale production of penicillin for use in World War II was undertaken in the United States, and it proved an incredibly effective antibiotic. A commemorative booklet has been prepared by the American Chemical Society and the Royal Society of Chemistry [1].

As a personal example of the unexpected from my own research laboratory, I was looking over some protein crystallisations set up around 10 years earlier by a graduate student of mine, Dr Suzanne

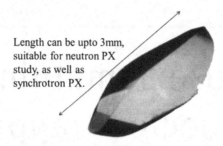

Length can be upto 3mm, suitable for neutron PX study, as well as synchrotron PX.

FIGURE 9.1   A crystal of the protein concanavalin A. The length can be up to 3 mm, suitable for neutron protein crystallography (PX) study, as well as synchrotron PX. This image is based on the research described in J. Habash et al. *Faraday Transactions* 1997, 93(24), 4313–4317. [3]

Weisgerber. The protein, isolated from jack beans, is called *concanavalin A*, and it binds sugar molecules. The crystallisations led to unusually large crystals, around 3 mm × 2 mm × 1 mm (i.e., 6 mm³; see Figure 9.1).

Although both have undergone major technological advances in recent decades, neutron sources still generate much weaker beams than X-ray sources. To produce diffracted beams intense enough to measure accurately, the neutron beam has to interact with a large (often unfeasibly large) mass of specimen. Fortunately, these crystals were so large that I knew a neutron crystal structure analysis was possible. Most crystal structure analyses use X-rays, but challenges to understand the function of a protein can specifically involve needing to know about their chemically active hydrogen atoms. A hydrogen atom, with its one electron, is much more difficult to see with X-rays than with neutrons. Because neutrons interact with nuclei rather than electrons, hydrogens compete with other elements on a more equal basis in terms of their scattering strength, especially if the hydrogens can be exchanged for deuteriums.

I set about collaborating with the scientists at the nuclear research reactor in Grenoble, for which the United Kingdom had a one-third share, to undertake the neutron diffraction data collection. Neutrons have a second very important property, besides allowing clearer visualisation of deuterium atoms: They do not cause radiation damage to the sample. In the late 1980s, I published a paper on the specific X-ray damage causing the splitting of the disulphide covalent bond in a protein [2]. Here, then, with neutrons, was a way of avoiding such X-ray damage. A further aspect was that research that I had been doing using the polychromatic X-ray emission from the synchrotron also could be applied to the neutron reactor emission of neutrons. This approach would increase the effectiveness of the use of the reactor's neutrons and in fact reduce the need for such large-volume crystals. A highly effective collaboration developed between my laboratory in Manchester and the scientists in Grenoble. In fact, the applications to many other proteins of such neutron diffraction studies are ongoing, and a former PhD student of mine, Dr Matthew Blakeley, is in charge of two of the Laue diffractometer (LADI) instruments, LADI-A and LADI-B. A new neutron source, the European Spallation Source, is also now under construction in Lund, Sweden. I am chair of the Neutron Macromolecular Crystallography Scientific and Technical Advisory Panel.

REFERENCES

1. Alexander Fleming Laboratory Museum. *The Discovery and Development of Penicillin, 1928–1945*. Accessed August 8, 2018, from https://www.acs.org/content/dam/acsorg/education/whatischemistry/landmarks/flemingpenicillin/the-discovery-and-development-of-penicillin-commemorative-booklet.pdf.
2. J. R. Helliwell. (1988). "Protein crystal perfection and the nature of radiation damage." *J. Crystal Growth* 90, 259–272.
3. J. Habash et al. (1997). "Neutron Laue diffraction study of concanavalin A: The proton of Asp28." *Faraday Transactions* 93(24), 4313–4317.

CHAPTER 10

Because Sometimes We Have to Interrupt a Line of Research Investigation

WHY DO SCIENTISTS SOMETIMES abandon research into a topic? It happens because we have hit some kind of limit. This can be the limit of current community research knowledge, one's own scientific limit of understanding, or limits in the resources required to take something further, preventing research from moving forward effectively. Or there might be another case—one that can be easily imagined—when a research grant proposal, no matter how well it is scored by a judging committee, simply is too far down on the ranking order of proposals to gain the necessary funding. Research grant proposal success rates vary quite a bit, but I have never known them to be better than 1 in 3, and they can be as poor as 1 in 8. Usually, as the principal investigator, one tries to bring forward such unfunded research at some level, but of course, it will need to have a reduced scope and the timetable must be longer.

The "limits of current knowledge" reason for not moving forward, applying either to one's own understanding or to the research community, is simply that sometimes the idea being advanced is ahead of its time. If so, one just has to hang onto it until it is ripe before trying again to move forward with it.

An alternative situation is when you complete research and development of a topic, but it needs an appropriate challenge to apply it to. This is especially true of research methods. An example of mine is where I completed the research and development on the method some 20 years before as an instruments scientist at the UK's Synchrotron Radiation Source. Then my research laboratory at the University of Manchester applied it to an actual study [1]. I gave an overview description here in reference [2].

REFERENCES

1. M. Cianci, P. J. Rizkallah, A. Olczak, J. Raftery, N. E. Chayen, P. F. Zagalsky, J. R. Helliwell. (2001). "Structure of apocrustacyanin A1 using softer X-rays." *Acta Crystallographica Section D-Biological Crystallography* D57, 1219–1229.
2. J. R. Helliwell. (2004). "Overview and new developments in softer X-ray ($2Å <λ< 5Å$) protein crystallography." *Journal of Synchrotron Radiation* 1, 1–3.

CHAPTER 11

Because We Want to Tackle Adventurous Research

A<small>T ANY STAGE OF</small> one's career, ideas can arise that are adventurous. This means that an idea might be totally unexpected, even appearing to oneself as being unlikely to work. In such cases, it may be necessary to explore it without research grant funding. One does not wish to share such ideas with a funding committee, not so much for fear of being labelled wild, but because the time that it takes to craft a proposal to perfection, with a typical success rate of between 1 in 3 and 1 in 8, is much more effort than simply trying it out. However, that means proceeding with little to no funding. But that is never entirely true because it is highly likely that one is working in a research environment, with access to the equipment that one might need. Also, if one's research is dependent on computational analysis, the computing power available on one's laptop is considerable, and basically then no extra funding is needed to undertake a study.

The issue likely has more to do with the cost of consumables (e.g., chemical or biological or material science). Of course, if your research is theoretical, then the question is how much time one wants to commit to it rather than to something more standard, more likely to lead to guaranteed results. Occasionally, funding agencies come to the rescue, as they can recognise that their schemes for funding tend to be directed to the well-accepted, next-step type of proposals: incremental science. Of course, incremental science is an incredibly powerful methodology.

As a simple example of its power, consider the iPad, a combination of several different functions, a considerable leap in capability over previous devices. There were undoubtedly innovations in the individual functions or sizes of the touch-sensitive screen. But all together, the functions that the iPad contains make it overall a remarkable device. But let's return to the topic of adventurous research. The Wellcome Trust ran a funding scheme for awhile called exactly that: "Adventurous Research Proposals." With a colleague, we submitted a proposal to the Trust. Because our audience was the Wellcome Trust, our proposal had to have a biomedical theme. We proposed to investigate protein structure when fully deuterated versus the usual protonated form [1]. The success rate was 1 in 30, but ours was one of the lucky few that got funded.

Obviously, the Wellcome Trust's plan was popular for the scale of funds committed to it, as judged by the large number of submissions compared to the funds available. There are other schemes offered by other institutions, usually for early career researchers, which provide seed funding so as to establish a line of research and allow the scientists to prepare a full application. Such programs are obviously a very good thing. Adventurous ideas can occur to scientists at all stages of their careers, even the semiretired scientist like myself.

REFERENCE

1. S. J. Cooper, D. Brockwell, J. Raftery et al. (1998). "The X-ray crystal structures of perdeuteriated and protiated enzyme elongation factor Tu are very similar." *J. Chem. Soc. Chem. Comm.* 10, 1063–1064.

II

The Role of Technology

CHAPTER 12

Because of Technology Push

E XPERIMENTAL SCIENCE WORKS WITHIN a framework of the current technology or what is coming soon. Ideas in science are called *science pull*, but developments in technology affecting science are called *technology push*. For example, the development of the performance of computer hardware over the decades has been remarkable. It is worth remembering that computers are a relatively recent innovation. They were invented during World War II with the Colossus project at Bletchley Park, and subsequently in peacetime came the first stored computer program, Small-scale Experimental Machine, nicknamed "Baby," at the University of Manchester in 1948. Indeed, how to write software, or rather the principles involved in programming a computer, were laid out by Alan Turing in 1936 [1].

The whole area of computer-based processing of experimental data and the calculation of computer simulations to predict new areas of science represent an incredible advance over calculations based on using slide rules or mathematical tables. The hardware has shrunk enormously, from a whole room of computer hardware to a laptop. This is all an example of technology pushing the limits of scientific opportunity. An aspect of this is how data storage capacity has been transformed over several decades. The "Baby" computer had a 32-bit word length and a memory of 32 words (1 kilobit). It occupied a room. On the other hand, my laptop now has 8 gigabytes of memory, along with 500 gigabytes of disk storage, and it occupies a small area of my desk.

The storage of data, which is much more readily gathered in digital form in the first place, can be part of large data stores offered commercially via cloud services, and at quite cheap rates. This has led to opportunities with so-called big data, where analyses can be made for trends or "needles in haystacks" in such large data sets. The opportunities with big data obviously are not only scientific applications, but also can include commercial and government activities. In my own research field of crystallography, we have always had a tradition of preserving the data underpinning our science articles. Initially, this was on paper or photographs, but of course, this process was transformed with the computer and disk storage. So started our crystal structure database, the Cambridge Structure Database in 1967, which then launched the Protein Data Bank jointly with Brookhaven National Laboratory in Upton, New York, in 1971. Each data-set entry was solely the atomic coordinates of a molecule derived from the crystal structure analysis, but as disk store expanded, the processed diffraction data measured on the experimental apparatus could be preserved too.

This store of data entries presented the readers of a publication with the opportunity to repeat many of the calculations of the original researchers. Most recently in my field, the International Union of Crystallography instigated a project called the Diffraction Data Deposition Working Group, which I chaired, to investigate the opportunities to preserve the much larger, experimental, raw data sets. This we found to be increasingly feasible, and there is now a range of opportunities for saving each raw data set with a crystal structure and its publication and database entry.

There are various categories of our research where this is most worthwhile, such as studies of molecular structure dynamics, which can violate the perfect layout of an ideal crystal, and the experimental signal appears outside the usual areas of the diffraction pattern that we measure. Another application is when there is more than one crystal type in a sample, and so the primary aim of researchers is to process the predominant experimental signal; however, the minor components of the sample also could be of major interest. This is the crystallographers' version of big data opportunities.

The technology push of the developing computer hardware capabilities opened up these new scientific opportunities for my research community, and the same is true for other research communities as well. Some communities acted as quickly as crystallographers to harness these new data technologies, including astronomers, particle physicists, and genomicists. Technology push is therefore a major driver for changing the scope of many areas of science. I will return to this theme later in Chapter 19 when I describe the "open science" concept, whereby if we can store raw experimental data, then we can ask the question, "What if we share those data as soon as they are measured with other scientists to bring about more rapid science discovery?" The urgent need for finding cures for serious diseases is a good example of such a driver.

The discipline of computer software has expanded enormously, such that it is now a vital aspect of science. Certainly in my field, software has become in effect a discipline in its own right. Computer science and mathematician specialists write marvellous code, and with it easy-to-use graphical user interfaces (GUIs). For more than a decade, I worked on learning computer programming languages, first BASIC and then FORTRAN. I wrote my own computer programs and taught a FORTRAN course when I was a lecturer in physics at the University of York.

Steadily, the specialists became predominant, and collaborative computer packages also emerged. Sharing of software code and open software code to facilitate progress in science became widely (albeit not completely) accepted. The code in our programming language that we can read and understand is compiled into a binary code that the computer recognises. Some software writers do not release their code, but of course, they like to see scientists use their software binary code in calculations; they wish to preserve the integrity of their code without variants being developed. (Another approach is to not put any comments in software, which makes it very difficult for people other than the author to understand it.)

So, like instrumentation and techniques, software underpins vast areas of modern-day science—indeed, probably all of it. If we want to know about how science was before the computer, we can turn to books on the history of science. The inventions of the first telescope and the first microscope revolutionised astronomy and biology, respectively. The computer, and the software that is run on a computer, is a technology that has revolutionised all of science, as well as everyday life.

REFERENCE

1. A. M. Turing. (1936–37). "On computable numbers, with an application to the Entscheidungsproblem." *Proceedings of the London Mathematical Society*, Series 2, 42, 230–265.

III

The Wider Research and Work Environment

Because We Wish to Engage in a Grand Challenge or Mission-Led Research Objective

ALMOST BY DEFINITION, GRAND challenges are so large in the scope and breadth of expertise needed to tackle them that large teams with diverse membership are needed to collaborate and undertake the research involved. A recent example of such a situation is the Global Challenges Research Fund (GCRF), connected with the UK Research and Innovation (UKRI) funding organisation. This has an assigned budget of £1.5 billion. Its aims are as follows [1]:

> To support cutting-edge research that addresses the challenges faced by developing countries. Alongside

the other GCRF delivery partners, we are creating complementary programmes that:

- Promote challenge-led disciplinary and interdisciplinary research, including the participation of researchers who may not previously have considered the applicability of their work to development issues

- Strengthen capacity for research, innovation and knowledge exchange in the UK and developing countries through partnership with excellent UK research and researchers

- Provide an agile response to emergencies where there is an urgent research need.

This type of fund, of course, is a directed one; that is, the themes for the research are handed down rather than coming up from individual researchers. That does not mean that the scientist's curiosity-driven aims cannot be met by joining in with applying for research funds within such an endeavour. Inevitably, though, there will be constraints and restraints on scientists during their participation in such a project, set by its scope and commitments. In any case, to survive in research, one needs at least one funded project—and what could be better than a directed one with such fine aspirations as this one from UKRI? [1]

REFERENCE

1. UK Research and Innovation (UKRI). *Global Challenges Research Fund*. Accessed August 8, 2018, from https://www.ukri.org/research/global-challenges-research-fund/.

IV

The Scientist's Inner Self

CHAPTER 14

Because We Wish to Develop Our Skills for a Better Future

A SCIENTIST'S SKILLS ARE VERY precious, whether they involve doing research or directing staff in their laboratory. For the former, the point is obvious because competently undertaking research requires properly learned skills. For a laboratory director, it is a balance of knowing what you direct your staff to undertake is feasible, as opposed to trusting them to do it and even to tackle something outrageously adventurous.

A treasured aspect of academic life is the research sabbatical. This is earned on a basis where usually 1 year in 7 can be expected to be research leave from one's duties of teaching and administration. So in my own academic career, I have taken research sabbaticals twice: in 1994 at Cornell University and the Cornell High Energy Synchrotron Source (CHESS) and also at the University of Chicago. In 2007, I took my sabbatical locally, at the University of Manchester, to learn (actually relearn—that is, bring myself up to date) on electron microscopy applied to determining the structure of a very large protein carotenoid complex (crustacyanin [1]). Part of my career was as a scientific civil servant, so I did not accrue time credits toward a sabbatical during those years.

Short visits are also practical at any time in a researcher's scientific career. For example, in 1983, I went to the University of Birmingham–Alabama to the Comprehensive Cancer Research Center Laboratory of Professor Charles E. Bugg ("Charlie") for one month with my wife and baby son. I had joined in a collaboration on an enzyme crystallography study, that of purine nucleoside phosphorylase. This has become one of my most cited publications [2]. It also led to a science popularisation article written by Charlie in *Scientific American* [3]. He cited my contribution specifically (namely, the work that I undertook at the Synchrotron Radiation Source in the United Kingdom). This was the instance where my mother and father, my aunts and uncles, and my cousins realised that I was doing something significant with my career.

REFERENCES

1. N. H. Rhys, M.-C. Wang, T. A. Jowitt, et al. (2011). "Deriving the ultrastructure of α-crustacyanin using lower-resolution structural and biophysical methods." *J. Synchrotron Rad.* 18, 79–83.
2. S. E. Ealick, S. A. Rule, D. C. Carter, et al. (1990). "Three-dimensional structure of human erythrocytic purine nucleoside phosphorylase at 3.2 Å resolution." *J. Biol. Chem.* 265, 1812–1820.
3. C. E. Bugg, W. M. Carson, J. A. Montgomery. (1993). "Drugs by design." *Scientific American*, December, 92–98.

CHAPTER 15

Because We Wish to Reach to an End Point

S CIENTISTS, LIKE MOST PEOPLE, like to see all aspects of a science research project properly explored. However, it is also said that "perfection is the enemy of progress." One can interpret this as meaning that the filling in of all the details takes a lot of effort and time but yields a decreasing impact compared with the core results. It is actually quite hard to apply this as a rule to one's scientific research planning. Nevertheless, it seems related to the Pareto principle of economics known as the 80/20 rule.

One source describes the Pareto principle can be applied to science, saying that "[t]he more predictions a theory makes, the greater the chance is of some of them being cheaply testable. Modifications of existing theories make many fewer new unique predictions, increasing the risk that the few predictions remaining will be very expensive to test. [1]" A self-regulating factor is whether one obtains the funding to undertake working out the remaining details of a research topic or theme. For instance, a funding committee and its chosen referees will quickly reject a project proposal that is already worked out rather than another that has much more useful or interesting discoveries to be made with it, but is not yet completely planned. Of course, in any one funding round there are always too many projects to fund, and in the end, reliance must be placed on the overall score attributed to a proposal. The ranking order based on such scores and the available funding determines where the funding cut-off is applied. It is inevitable that the scores will have some degree of error in their estimation, and some proposers are lucky whilst others are not. Therefore, it is always better to be at the start of a research theme rather than at the end. The start, however, will be the most innovative moment of a research project and is arguably the most vulnerable stage for determining whether one can attract funding for it or not, as I describe in Chapter 8 of this book.

REFERENCE

1. Pareto, V. (1896). *Cours d'économie politique*. Published by F. Rouge, Lausanne.

CHAPTER 16

Because We Like Finding Things Out "At the Science Bench"

A T THE START OF a scientist's career, as a PhD student, one is at the bench doing research, of course. That work bench may be in the wet lab, at the computer, or with a pencil and paper writing equations. As a scientist's career progresses, it likely will lead to the running of a research group to a greater or lesser degree. At the bare minimum, a research group will include PhD students to be supervised and trained. In addition, there will usually be postdoctoral research assistants working on these funded research projects.

At this point, some scientists can be put in charge of research groups that are so large that the supervision of them requires one or even a layer of "middle manager" scientist positions, and the lead scientist will likely be solely involved in administration.

I found this stage particularly tense because I found it impossible to be directly at the laboratory bench myself on a daily basis and to run a large team at the same time. Other scientists are quite happy with such a situation. So the title of this chapter is admittedly overly simplistic, but it nevertheless reflects my own preferences.

V

Communication of Science

CHAPTER 17

Why Do Scientists Confer So Much?

A T MY LOCAL TENNIS club, I am regularly asked where I am going next for my work trips. It sounds so glamorous to my friends that I head to distant places. They don't quite believe that I would spend so much time in lecture halls when visiting exciting-sounding locations. They rarely ask what my conference is about for that next trip. They are actually very interested in science, but mainly mysteries such as "What is dark matter?" or "Why or how does the universe come into being?" More basically, they see the fruits of science and technology advances around them, and they greatly respect science.

So, why do we as scientists go to conferences and be away from family and home so much, and away from the laboratory as well? First and foremost, we are eager to present our results and to listen to the best results from other laboratories. There also may be business meetings to attend with one's research community's organisation and, naturally, committees. Most conferences are packed with lectures and meetings morning, afternoon and evening. They are actually quite exhausting.

One has to be exceedingly careful when presenting one's results. Most simply, one rule is to stick to the stipulated time for speaking and thereby ensure that there is time to field the audience's questions. This is the feedback that one seeks on the results. Strangely, there will always be lecturers who run over time and use up their allowance of time for questions. Even more serious is to know when a piece of research is ready to be presented. My ideal time is after acceptance for publication, but before open release. One has the peace of mind then that all likely criticisms from referees have been aired and dealt with in article revisions. If the research has patent possibilities, then one simply does not present the work until the Intellectual Property Department of one's university has been consulted.

Around the lectures and committee meetings, numerous discussions go on, both informal and semiformal. In such sessions, joint research proposals are planned for collaborations, or job hunting by researchers who seek to meet their possible future laboratory head.

Can conferences be avoided? Well, of course. Indeed, remote access to a conference via the Internet might be possible, and in special circumstances a lecture can be presented from a remote location into the conference. Asking questions from a remote location is probably the most awkward experience. It is much better to be directly present in the lecture hall, not least for the more detailed discussions that may ensue in the coffee or tea breaks during the formal meetings.

An excellent illustration of a conference type aimed at maximising discussion is the Faraday Discussion series run by the Royal Society of Chemistry. A recent example, which I helped organise, is described in Reference 1.

Not only does one attend many conferences, when one adds them up through one's career, but in my case, I served on many organising committees for them. I also chaired large and small conferences and workshops. Examples included chairing the Local Organising Committee for the British Crystallographic Association

Annual Conference in 1993, held in Manchester; and chairing the Programme Committee for the British Crystallographic Association Annual Conference, held in York in 2015. Those two are conferences with between 200 and 350 persons in attendance. In 2002, I chaired the Faraday Discussion on Time-Resolved Chemistry held in Manchester; this was jointly with a colleague, Professor Chick Wilson.

I have not run a full conference in Europe or an International Union of Crystallography (IUCr) World Congress, which are truly massive logistical operations with between 1,000 and 2,000 persons attending. But I have chaired portions of such conferences, either satellite meetings or workshops [for example, in Beijing in 1993 at the IUCr World Congress on Synchrotron Radiation Crystallography and in 2018 at the European Crystallography Meeting (ECM31) in Oviedo, Spain on Neutron Macromolecular Crystallography.]

REFERENCE

1. J. Segarra-Martı, R. Ramakrishnan, J. Vinals, and A.J. Hughes. (2018). *Highlights from the Faraday Discussion on Photoinduced Processes in Nucleic Acids and Proteins.* Doi: 10.1039/c8cc90123f.

CHAPTER 18

Why Do Scientists Submit Their Research to a Journal?

A REPUTABLE JOURNAL IS ONE that organises the refereeing of a submitted research article in a thorough and well-informed way. This requires an experienced editor consulting expert referees (usually two or more). The editor is the person who finally accepts the submitted research article as fit for publication. The track record in doing this well is that a journal's articles have an impact and that is long lasting.

Publishers of various types exist. Commercial publishers have sought out a profit-based approach, and profits of around 40 percent are there to be seen and wondered at. Learned societies, as they are called, are the professional science associations with qualified scientists as members, and who also run journals and make relatively small profits, which are used to support their research communities. Either way, in a commercial or learned society, publishers provide a service because they do not fund the scientific research that is undertaken. Naturally, funding agencies have started to try to reshape this process in order to control and reduce these profits, as well as to avoid a paywall barrier to the public, who might wish to read their research publications. It is the public, after all, who fund much of the research in the first place through their taxes. This "open access for readers" revolution is forcing changes in how scientists publish. Scientists submit articles to the peer-review process before publication because they seek the scrutiny of their work by expert peers, with their recommended changes being made or refuted by the authors.

New models of publication are being used. These mainly revolve around the preprint approach and have been the methods adopted by the physics community for many years, notably via its arXiv preprint archive [1]. This allows a wide community readership of an article prior to submission to a journal, and feedback to the authors is made directly. The submitted article is quite possibly different from the preprint version. The final publication is a further revised version following scrutiny by a journal's referees and its editor. There is also the postpublication peer review commentary, as a critique of a publication can be published by others along with the original authors' response.

An overall disadvantage of a journal's procedures is that the reports of its referees are anonymous and also very likely are lost to the science record. Here too, changes are occurring where a journal such as *eLife* is publishing the referees' reports as well as the editor's summary decision [2].

Another approach is that being pioneered by the Faculty1000 project (F1000) [3]. This is an author-driven publishing model that offers immediate publication, the advantage of the preprint server, with transparent and open peer review [3]. The unusual aspect here is that the authors choose peer reviewers rather than having an editor look over their work. That the peer reviewers and their reports are not anonymous is the counterpoint to the natural worry that the readers may have that the peer reviewers might be biased in favour of the authors. Conversely, this approach avoids any bias that an editor may have. There is an editorial team who serve as intermediaries between authors and potential peer reviewers. At the time of writing this book, F1000 has published more than 2,000 articles [4]. It is also possible to post comments on the article and or reviewers' reports by any reader.

A further implementation example of this is by the African Academy of Sciences (AAS), using the tools provided by F1000 [5]. The first batch of articles has recently been published by the AAS, for which one has a referee's report and can thereby serve as an example of this style [6]. The AAS is working with the Accelerating Excellence in Science in Africa program, which funds projects; financial supporters includes the Bill and Melinda Gates Foundation, the Wellcome Trust, and the British government's Department for International Development (DFID) [7]. This is a very fine initiative.

Another new vision for change is also surfacing, which is known as the *blockchain*, sometimes also referred to as *web 3.0*. This seeks to document the entire linked chain of science ideas, from proposal, to research undertaken, and finally to publication. This total documentation of research metadata is envisioned to proceed with digital tokens of value being awarded throughout. This would include tokens being awarded to referees. This would be a new evolution of the process and methodology of science, maybe even a revolution. For further details on this, see [8]. It is also linked with efforts to make an open science environment that harnesses communitywide efforts to bring about more rapid research solutions, such as in response to urgent disease challenges.

REFERENCES

1. arXiv Cornell University Library (https://arxiv.org/).
2. *eLife*. IRF4 Haploinsufficiency in a Family with Whipple's Disease. Accessed August 9, 2018, from https://elifesciences.org/articles/32340.

3. F1000 Research. (https://f1000research.com/).

4. S. A. Frank. (2018). "A biochemical logarithmic sensor with broad dynamic range. F1000 Research. Accessed August 9, 2018, from https://f1000research.com/articles/7-200/v3.

5. AAS Open Research (https://aasopenresearch.org/contact).

6. A. E. Ahmed, P. T. Mpangase, S. Panji, et al. (2018). "Organizing and running bioinformatics hackathons within Africa: The H3ABioNet cloud computing experience." AAS Open Research. Accessed August 9, 2018, from https://aasopenresearch.org/articles/1-9/v1.

7. *AAS Launches Alliance for Accelerating Excellence in Science in Africa.* Video. Accessed August 9, 2018, from https://www.youtube.com/watch?v=JXPps8xsvjM.

8. S. Bartling author page. Blockchain for Science. Accessed August 9, 2018, from https://www.blockchainforscience.com/author/soenkeba/.

VI

Science and Society

Because We Can Expand the Scope of Research with "Open Science," Bringing an Improved Future for Society

T HE TERM *OPEN SCIENCE* is recent. It has arisen as a policy push of the funding agencies to realise science advances quicker when meeting directed science challenges set by the agencies themselves. The urgency often arises from a health or disease emergency that requires concerted research by the relevant scientific community to find possible solutions rapidly. Of course, concerted efforts are in themselves not new. World War II brought concerted efforts to bring about large-scale production of the antibiotic penicillin. Open science today arises in the frame of being able to share raw scientific data more or less as soon as they are measured (i.e., well before publication). Thus, large teams can be in contact even whether they are on opposite sides of the world. The Internet, invented at the European Organization for Nuclear Research (CERN) in Geneva, is key to this approach. The particle physicists saw the need for both remote access to the CERN particle accelerators and for the sharing of data between collaborating research groups. Such an approach is now applicable to emergency disease challenges. The sense here is that there should be openness between collaborating research groups that may ordinarily be competitors. That does not necessarily mean that the raw experimental data are immediately open to the public. Another version of open science is what is termed *citizen science,* where members of the public can lodge their own observations at a centralised website. Thus, large-scale, geographically disparate observations can be gathered.

Open science also can (indeed should) be a key feature of regular scientific research. The results of publicly funded science should be openly available to the taxpayers who funded it. This is termed *open access.* The data that underpin research publications funded by taxpayers also should be openly available. Published results, of course, appear at the end of a research project or an identifiable milestone of a larger research project. Such open science is with a research team working on its own, whereas the previous discussion focussed on the sharing of raw scientific data as soon as it is measured to allow very broadly based collaborations to develop.

There are interesting exceptions to the open science approach. Closed science, as one might call it, is not automatically bad. One can cite several generic examples. Single individuals working alone can undertake research stemming from their own imagination. In the early stages, this is very likely to be unfunded. Results can be written up and submitted for publication. Often, these studies can be the most innovative and unusual. There are no monies, however, for research assistants or for publication fees to make the publications open access.

This emphasises the need for researchers to keep their skills up to date so that they can work on their own on such ideas. It also illustrates that subscribers to journals cover the publication costs of such unfunded research, although it does put the research publication behind a paywall. Some journals have been established by research communities (namely, the learned societies). Such is the case with my own research area of crystallography, as the International Union of Crystallography runs its own journals and is a not-for-profit science publisher. Another approach is that of the Wellcome Trust, which has its own journal called *eLife*. It finances the journal completely, without subscriber fees or article processing charges; it is free to readers and free to authors.

Open science is practical in the situations that I describe here, which we as scientists engage with wherever possible.

CHAPTER 20

Because a New or Improved Product Is Needed via Industrial Research

WHEN I WORKED AS a scientific civil servant at the Synchrotron Radiation Source in the United Kingdom, I costed out the use of beamlines for protein crystallography by industry. The companies paid for their usage time, approximately £2000 per 8-hour shift, and their researchers would collect X-ray diffraction data from a protein that they had crystallised and was a company drug target. Typically, their crystals were grown with an organic molecule bound to the protein active site, and the results allowed them to see how they might optimise that compound to be a stronger binder and so block an important function of the protein in, say, a bacterium. There was no obligation on the company to publish, as it had paid for its beamtime, unlike the academic users. These pursuits fell within a pharmaceutical company's research and development activities. Industrial research had the need for such leading analytical research services, which we provided. There were similar services to that of crystallography involving X-ray spectroscopy and powder X-ray diffraction, which were applied to characterising other products, such as a company's search for a new drug polymorph (perhaps for the purpose of getting around a patent held by another company, or simply a more effective pill version of its own product). Sales of medicines can reach the billions of pounds annually. Thus, our analytical services at the very modest rate of £2000 per 8-hour shift were a tiny, tiny fraction of that.

Traditionally, pure research has been undertaken in universities and applied research has been undertaken in industry. This boundary is much less clear today. Universities encourage their staff to be on the lookout for commercial application possibilities of their pure research. Support teams in university administrations are available to the academic researcher. They are familiar with patenting and with launching spinouts of new companies from university research. Academics who are successful with this sort of research describe new ideas and innovations bouncing back and forth between university laboratories and company laboratories [1]. Funding agencies have different rules for the commercialisation of pure research, so careful auditing and even physical separation of laboratory bench space are required.

The most successful, and wealthiest, companies today are not only pharmaceutical companies. It is easy to notice the vast wealth accumulated by Apple from sales of its iPhone and iPad, which are incredibly popular around the world. Each new generation of these devices brings an innovation of some sort, based on research and development in the company, such as fingerprint, voice, or facial recognition. Fundamental science projects form the foundation of these initiatives, but bringing them together to create devices that work and are saleable is the domain of industry.

An example of the close interest that government takes in linking science and innovation with industry is the Government Office for Science, whose priorities are [2]:

- Supporting national growth and increasing the UK's productivity by linking science, innovation and industrial enterprise

- Supporting regional growth by building on existing science and innovation activity across the country

- Using technology to develop modern and cheaper public services

- Preventing or addressing emergencies and mapping national security risks

and which states its responsibilities as follows:

- Giving scientific advice to the Prime Minister and members of the Cabinet, through a programme of projects that reflect the priorities of the *Government Chief Scientific Adviser*

- Ensuring and improving the quality and use of scientific evidence and advice in government (through advice and projects and by creating and supporting connections between officials and the scientific community)

- Providing the best scientific advice in the case of emergencies, through the *Scientific Advisory Group for Emergencies (SAGE)*

- Helping the independent *Council for Science and Technology* provide high-level advice to the Prime Minister

Naturally, one can be intrigued by why the Government Office for Science has no declared interest in promoting fundamental science. However, this is linked with the Haldane principle, introduced in

1918, which states that research should be separate from, and not be directed by, government, but whose direction is judged by academic researchers according to assessing proposals by criteria of excellence. Here is a description of the Haldane principle [3]:

> [the United Kingdom's] Secretary of State for Innovation, Universities and Skills put his mark on the Haldane Principle in April 2008:
>
> > For many years, the British government has been guided by the Haldane principle—that detailed decisions on how research money is spent are for the science community to make through the research councils.
> >
> > Our basis for funding research is also enshrined in the Science and Technology Act of 1965, which gives the Secretary of State power to direct the research councils—and, in practise, respects the spirit of the Haldane principle.
> >
> > In practise, of course, Haldane has been interpreted to a greater or lesser extent over the years, not least when Ted Heath transferred a quarter of research council funding to government departments—a move undone by Margaret Thatcher.
> >
> > But in the twenty-first century, three fundamental elements remain entirely valid.

- That researchers are best placed to determine detailed priorities.

- That the government's role is to set the over-arching strategy; and

- That the research councils are 'guardians of the independence of science'.

The money spent on research and development (R&D) in business enterprises is the largest fraction. The European Union

(EU), as well as individual governments, follows the funding of research patterns closely. For example, quoting the European Union:

> An analysis of R & D expenditure by source of funds shows that more than half (55.3%) of the total expenditure within the EU-28 in 2015 was funded by business enterprises, while almost one third (31.3%) was funded by government, and a further 10.8% from abroad (foreign funds). Funding by the higher education and private non-profit sectors was relatively small, 0.9% and 1.7% of the total respectively. These shares were relatively stable over time. [4]

This same report from the European Union declares a context [4]:

> Through its *innovation union flagship initiative* (which forms part of the Europe 2020 strategy) the *European Commission* has placed renewed emphasis on the conversion of Europe's scientific expertise into marketable products and services, through seeking to use public sector intervention to stimulate the private sector and to remove bottlenecks which stop such ideas reaching the market.

The amount of time that a researcher is expected to commit to supporting this laudable context set by the European Union is undefined. Naturally, only individuals can decide what they believe are their priorities for the best research that they can do, but suffice to say, science researchers are in a complicated landscape.

REFERENCES

1. Blundell, T. L. (2017). Protein crystallography and drug discovery: Recollections of knowledge exchange between academia and industry. *IUCrJ* 4, 308–321.

2. Government Office for Science. *About Us.* Accessed August 10, 2018, from https://www.gov.uk/government/organisations/government-office-for-science/about.
3. British Parliament. *The Haldane Principle.* Accessed August 10, 2018, from https://publications.parliament.uk/pa/cm200809/cmselect/cmdius/168/16807.htm.
4. Eurostat. *R&D Expenditure.* Accessed August 10, 2018, from http://ec.europa.eu/eurostat/statistics-explained/index.php/R_&_D_expenditure.

CHAPTER 21

Importance of the "Whys of a Scientific Life" for Society

O NE OF THE GREAT successes of science is the change in life span of people over the recent past. This involved the coming together of a variety of improvements: in agricultural science, medical science and treatment (with clean water, new medicines, antibiotics, vaccination, and physics-based instruments for patient monitoring), and last but not least, improved knowledge of nutrition. To what extent the types of science described throughout this book contributed to these accomplishments can be difficult to disentangle. There are certainly identifiable important moments, such as Sir Alexander Fleming's discovery of penicillin and its subsequent production as an antibiotic. Largely though, it is through the systematic march of science development proceeding by increments in the various areas I have listed here. Certainly, the direction of U.S. president John F. Kennedy galvanised the National Aeronautics and Space Administration (NASA) to strive for that goal of landing astronauts on the Moon, which was spectacularly achieved. Perhaps a similar announcement to place a person on Mars will be achievable as well.

Eventually, though, we have to think that there can be a limit on such plans—when does science planning become indivisible from science fiction? In imagining our future, we must promote peace on Earth; scientists must get involved, and lead, as only we know what might be possible using science.

VII

Fundamentals Revisited

CHAPTER 22

Why Is Science Objective? Because of Data and Peer Review

IT CAN BE ARGUED that science is only minimised subjectivity, and that viewpoint, extrapolating further, allows for advocacy in science [1]. I have argued strongly against this [2]. The concept of minimised subjectivity was emphasised earlier than Reference 1 in Reference 3, but without extrapolating to advocacy. Why is this issue important? The public and government seek a coldly dispassionate view of facts from science advisors. I believe that we can achieve that ideal because of two aspects of the scientific research enterprise: peer review of article along with its underpinning experimental data, both before publication, by referees and other experts, and afterward by the readership. Furthermore, the collective knowledge of the underlying data as the foundation of individual publications is gathered in archives and databases. My logic means that submitted articles whose data have not been peer reviewed (and, if necessary, amended) before publication can be subjective. It is, therefore, vital that peer review of data together with the words of an article must be mandatory.

A related aspect of science being objective is that to qualify as new science, a publication must report a significant advance for the first time. This is not the domain of the authors, who can *claim* a significant advance, but this must be determined by three parties: the referees, the journal editors, and finally, the readers.

One metric of a science journal's objectivity and the significance of an article comprising science results is measured by the number of citations and amount of social media attention that a journal and an article can receive. A further measure of impact of some science results is that they find uses—such as consumers finding that an application works and has benefits. The World Wide Web works; the iPad and smartphones work; nuclear power works; and genetic fingerprinting works. These are objective measures by any person's criteria, whether scientist or layperson. Advocacy is not required to convince us of these facts.

Why is the saying "Take no one's word for it" important? It has been the motto of no less an organization than The Royal Society since it was founded more than 350 years ago (https://royalsociety.org/about-us/history/). A publication's words, then, present a narrative, whereas the underpinning data are its foundation. However, such data also can lend themselves to other narratives by further scientific insights and interpretations. The International Council for Science has a dedicated Committee on Data (CODATA), which has helped develop a policy for all scientific data—namely, that it be *FAIR*, which means findable, accessible, interoperable, and reusable. (For an example of such a scenario, see https://dspacecris.eurocris.org/bitstream/11366/606/1/euroCRIS_SMM_Bratislava_SHodson_FAIRdata_CODATA_20171121.pdf.) However, there are situations where scientists can have legitimate differences of opinion on presented data that can be significant. A simple example is when the statement in an article that some data sit on a curve rather than a straight line is not accepted by referees. An editor of a science journal could acknowledge this critique by concluding that the article's authors should measure more data to resolve the interpretation of the data as a line or a curve.

In another example, the controversies surrounding climate change have arisen due to small effects, where measurement errors, or insufficient time periods to establish trends firmly, have to be dealt with clearly. But these controversies sometimes also are pushed by advocates of partisan industries, and politicians of countries that benefit greatly from those industries. They seem to be acting simply to protect their future profits and economies. They fear that acknowledging climate change is actually happening means that they must agree to changing their industrial methods. So this can foment great tension between "objective scientists" and "partisan politicians."

REFERENCES

1. Bradshaw, C. J. A. (2018). The Effective Scientist A Handy Guide to a Successful Academic Career CUP.
2. Helliwell, J. R. (2018). *J Appl Cryst* 51, 1259–1261. Book review of Corey J. A. Bradshaw. The Effective Scientist—A Handy Guide to a Successful Academic Career CUP 2018.
3. Ziman, J. (2000). Real Science: What it is and What it means. CUP.

CHAPTER 23

Why Is Science a Joy to Do?

MOST SCIENTISTS ENJOY GATHERING data in an experiment. There most often is a theory (or at least a hypothesis) behind it. The first step toward interpreting experimental results is to plot the data points. Next comes the interpretation of them, such as the question of whether the information should be presented as a straight line (whether a dependent variable y varies as x, the independent variable), a curve ($y = x^2$), or an exponential change ($y = e^x$).

This statement of these simple steps captures much of the joys of a scientific life, which is one of the fundamentals of why people become scientists—experiencing joy. As founding chairman of the European Crystallographic Association's Instruments and Experimental Techniques Special Interest Group (the IET SIG), my own joy comes from measuring the best data, having decided which data should be measured. This is apparent in the following simple example. For y, I would pick an atom's movement and for x, a parameter such as time when I have placed an enzyme molecule as a chemical catalyst with a reactant that helps make a product.

The precision with which I can locate that atom would not be perfect, and separating the error of measurement from the atom's contribution to the catalytic process is a challenge to me as a trained physicist so as to reveal the chemistry. To set up ideal chemical conditions is the challenge to me as a trained chemist. To prepare a pure enzyme from a sheep's liver, for instance, is a challenge to me as a trained biochemist molecular biologist. This was the topic area that I picked for my research doctorate at Oxford University in the mid-1970s. Through the different parts of my career I have learned each of these science subjects. Chemical catalysis e.g. by an enzyme involves accelerating the rate of conversion of a molecular reactant to a different molecular product.

Throughout my career, I have found it a captivating challenge to maintain my skills as new technologies have come, and working at this helps me maintain the basic joy of science. Other scientists would be excited by other topics, but the basics of a graph of data points would be the same for all of us. An astronomer/cosmologist could plot the distance to stars, both near and far from Earth, on a graph versus time, and thus determine whether there is a steady expansion of the universe or an acceleration as well. Meanwhile, an economist could plot inflation versus time and decide if a country's economy were basically stable or inherently on the brink of a total societal breakdown, where its population would eventually be unable to afford to buy anything; a current example seems to be Venezuela, with its annual inflation of 14,000 percent and growing. A physicist could look at the deflection of light that passes close to a planetary body, such as can be monitored during a total solar eclipse, which is when the Moon passes between the Earth and the Sun. A biologist can plot the number of algae in a dish over an amount of time and decide if its population increases linearly or not, and also observe the impact of varying the food composition fed to the algae. As algae can be grown to be used for fuel, discovering the fastest possible growth time would be important.

These examples capture a mixture of basic and applied science. John Ziman's book *Real Science* [1] examines science and its progress from the point of view of epistemology, the study of knowledge itself, such that (location 1830 of 5571 in my Kindle for ipad copy) "the whole argument of this book is that there is no way of ensuring that we have got things right scientifically." Ziman states that (location 4149 of 5571 in my Kindle for ipad copy) "the point is very simple. . . academic science relies on the exercise of intersubjectivity, directed towards achieving consensus." This is a totally misleading statement in my view, as it leaves out the role of data, which I have emphasised in this discussion, as the foundation of science being objective rather than subjective. (It may be relevant to mention here that I am chair of the International Union of Crystallography's Committee on Data, and John Ziman served as convenor of the Epistemology Group of the Council of Science and Society in the United Kingdom.) These two perspectives emphasise the different views that we each have taken as professional scientists.

To finalise the description of my own worldview, scientific data are of direct interest and play a key role in science, in all societies across the globe, regardless of gender, beliefs, or languages. Ziman's "Consensus" as a concept of what is accepted as science is incorrect. A telling example is Albert Einstein's question, "What if the speed of light is finite, and what implications would that have?" which broke the consensus of how physicists viewed physics in 1905.

REFERENCE

1. J. Ziman. (2000). *Real Science: What It Is and What It Means.* Cambridge University Press, Cambridge, UK.

Appendix: Why Do Some Scientists Write How-to Guides about Science?

I HAVE CHARACTERIZED THIS DISCUSSION an appendix because only a few scientists are motivated to explain how-to guides, but some of these books are about why scientists do what they do.

In the Bibliography, I collate several examples of these books; I found the order of each author's chapters especially interesting, as they reveal some variety of emphases as well as commonalities. I would note some interesting points. Selye (1964) uses extensive descriptions of his research on the physiology of stress experiments on rats to illustrate his points; I did not find that effective because it was largely opaque to me (and, I confess, I am rather squeamish). Nancy Rothwell's book often mentions *On Being a Scientist*, from the National Academy of Sciences (1995). Barbara J Gabrys and Jane A Langdale (2011) book often cites *The 7 Habits of Highly Effective People.*

I believe that a strength of my book *Skills for a Scientific Life* is that it brings together the learning that I gained both as a scientific civil servant, including my work providing analytical services to industry, up to the director level, and as a professor and as a senior mentor for new academics. A notable strength of Gabrys and Langdale's book was their descriptions of workshops that they had run on different aspects of the scientific life. John Ziman's book *Real Science* assesses what science is and what it means, and he seems to especially like, by way of analogy with Darwin's theory of evolution, Blind Variation Selection Retention (BVSR). He provides an epistemological approach to the discussion. BVSR, Ziman asserts, involves the variations of different scientists' research being the 'blind variation' component of Darwinian evolution theory. He also asserts that retention is the survival of the fittest, selection, component of Darwinian evolution theory such that the wider scientific community retains those ideas and developments in science that work. Overall, I really like Peter Medawar's (1979) book *Advice to a Young Scientist*.

BIBLIOGRAPHY

K. Barker. (2010). *At the Helm: Leading Your Laboratory*, 2nd ed. New York: Cold Spring Harbour Press.

C. J. A. Bradshaw. (2018). *The Effective Scientist: A Handy Guide to a Successful Scientific Career*. Cambridge, UK: Cambridge University Press.

S. R. Covey. (2013). *The 7 Habits of Highly Effective People: Powerful Lessons in Personal Change*. New York: Rosetta Books.

B. J. Gabrys, and J. A. Langdale. (2011). *How to Succeed as a Scientist: From Postdoc to Professor*. Cambridge, UK: Cambridge University Press.

J. R. Helliwell. (2016). *Skills for a Scientific Life*. Boca Raton, FL: CRC Press, Taylor and Francis Group.

F. MacRitchie. (2011). *Scientific Research as a Career*. Boca Raton, FL: CRC Press.

P. B. Medawar. (1979). *Advice to a Young Scientist*. London: Penguin Books.

National Academy of Sciences. (1995). *On Being a Scientist: Responsible Conduct in Research*. Washington, DC: National Academy Press.

J. W. Niemantsverdriet, and J-K Felderhof. (2017). *Scientific Leadership*. De Gruyter, Berlin.

N. Rothwell. (2002). *Who Wants to be a Scientist? Choosing Science as a Career.* Cambridge, UK: Cambridge University Press.

H. Selye. (1977). *From Dream to Discovery. On Being a Scientist.* New York: Arno Press.

E. O. Wilson. (2013). *Letters to a Young Scientist.* New York: Liveright Publishing Corporation, a Division of W.W. Norton & Company.

J. Ziman. (2000). *Real Science: What It Is and What It Means.* Cambridge, UK: Cambridge University Press.

Index

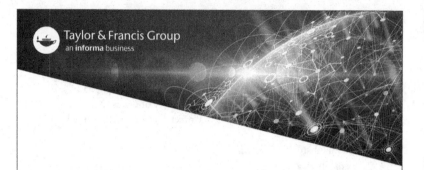